THE SEA LIBRARY

OCEAN PLANTS

by Sue Beauregard and Jill Fairchild

The Children's Book Company

Consultant: Heidi Hahn
Scripps Institution of Oceanography

Photography by Robert B. Evans: cover, 2–3, 8, 12, 31; Jim Cooluris: 1, 15; Jim and Cathy Church: 4; Henry C. Genthe: 7, 20; R. Trelease: 11; Steve Williams: 16; C. W. Sullivan: 19; Carl Roessler: 22, 26; James Scott: 25; Stan Keiser: 29.

Designed by Richard Crawford and David Mekelburg.

Published by Cypress Publishing Group,
Glendale, California 91204.
Distributed by The Children's Book Company,
Mankato, Minnesota 56001.
Printed in the United States of America.
Library of Congress Number: 77-71274
International Standard Book Number: 0-89813-026-3

Imagine you are underwater. Enjoy floating and swimming and looking around.

You see great tall ocean plants. You can touch the strong stalks and the long blades. The blades look like leaves, but they're not leaves. You feel the plants swaying as the water moves.

Look up. Do you see sunlight streaming through the water? The sun is important to plants.

4

There are gardens and forests in the ocean.

Plants of the ocean have many shapes, sizes, and colors. Most ocean plants are algae.

These green algae look like a cluster of glistening grapes.

Algae don't have roots or leaves or flowers or seeds, as most land plants do. But all plants have one important thing in common. All plants have green chlorophyll. Plants need chlorophyll. They use chlorophyll to make their own food.

As plants grow, they give off oxygen. All life on the land and in the sea needs oxygen. Without oxygen, there would be no life on the earth. Plants supply food and oxygen that all living things need.

All plants need water and nutrients and sunlight to grow.

In the ocean, getting enough sunlight is a problem. Different kinds of plants need different amounts of light. Plants that need the most light must grow near the water's surface where there is much sunlight.

These algae are called rock weed. They grow in shallow water where there is plenty of sunlight.

Ocean waves rush back and forth over the algae, but the plants are not washed away. They are attached firmly to rocks.

Brown algae called kelp grow in waters up to 30 meters deep.

There are many different kinds of kelp. The kelp you see here grows thin, long, and strong stalks. The stalks twist together as they grow. Blades grow off the stalks. The stalks carry their blades up to the sunlight. But the stalks are not stiff like tree trunks. When you turn the page, you will see what helps to hold them up straight.

Kelp grows golden round bladders filled with air. The air bladders work like balloons. They help to lift the great stalks of kelp to the surface.

Kelp is anchored to rocks on the ocean floor with a part of the plant called a holdfast. The holdfast isn't a root. It only anchors the plant.

Thousands and thousands and thousands of animals live in the giant kelp forests. Some small animals live inside the holdfasts. Some live on the blades. Other larger animals live and swim in and around the kelp stalks and blades. They all find food and protection there.

The most important plants in the ocean are tiny algae called plant plankton. They are so small you need a microscope to see them. They need plenty of sunlight. Their small size helps them to live near the ocean's surface.

Ninety-nine percent of the ocean's plants are plankton. They supply most of Earth's oxygen.

Plant plankton are called the "pasture" of the sea because they are the food for so many ocean animals.

Is this a fish or a plant? It's a fish.

The plant you see behind the fish is a brown algae called sargassum. It floats far out at sea. It is held up by small air bladders.

The sargassum fish lives in disguise in the sargassum. Its shape and color are almost the same as the algae.

Many kinds of animals you would expect to see only near shore live in the sargassum.

This barracuda is swimming above eel grass. Eel grass grows on sandy ocean floors in shallow waters. It is one of the few ocean plants that is not algae. Like all grass, eel grass grows roots, flowers, and seeds.

In the coral reefs, hanging veils of green algae grow among sponges and corals.

Many varieties of algae grow in shallow reef waters.

The green color of these corals comes from green algae. The algae live inside the corals. The plants are too tiny to see without a microscope. Many corals can't live without these algae. And the algae can't live without the corals. The algae supply oxygen for the corals as well as some of the food they need. The corals give protection and nutrients to the plants.

Many beautiful algae seem to glimmer and glow with neon colors.

The plant that looks like purple lettuce is called a red algae. Another kind of red algae is a hard flat-spreading algae. It looks like pink cement. It grows on the tops of rocks and old shells.

Ocean plants give beauty to the underwater world. They provide food and shelter for ocean animals. More than that, they give oxygen—the air we breathe.

Take a closer look: Name what you see here.
Match the photographs to the pictures inside your book.